@doodlesbydrewski

COPYRIGHT © 2020
DOODLES BY DREWSKI

ALL RIGHTS RESERVED,
INCLUDING THE
REPRODUCTION
IN WHOLE OR IN
PART IN ANY FORM

HEART-LESS

doodles by drewski

BLOBBY FLAY

doodles by drewski

DEAD A HEAD

doodles by drewski

SOLID WASTE

doodles by drewski

FOUR POP

doodles by drewski

HAND STAN

doodles by drewski

doodles by drewski

NACHO POTATOES

doodles by drewski

NERVIS CENTER

doodles by drewski

DUCKIT

10.

doodles by drewski

HEAD & SHOULDERS 11.

doodles by drewski

3 BLIND PIGS

doodles by drewski

MIND BUG

13.

doodles by drewski

KOI BOI'S   14.

doodles by drewski

UNCLE FISTER

15.

doodles by drewski

EGGS ON TOAST 16.

doodles by drewski

ROOTY FROOTY 17.

doodles by drewski

GROLL

18.

doodles by drewski

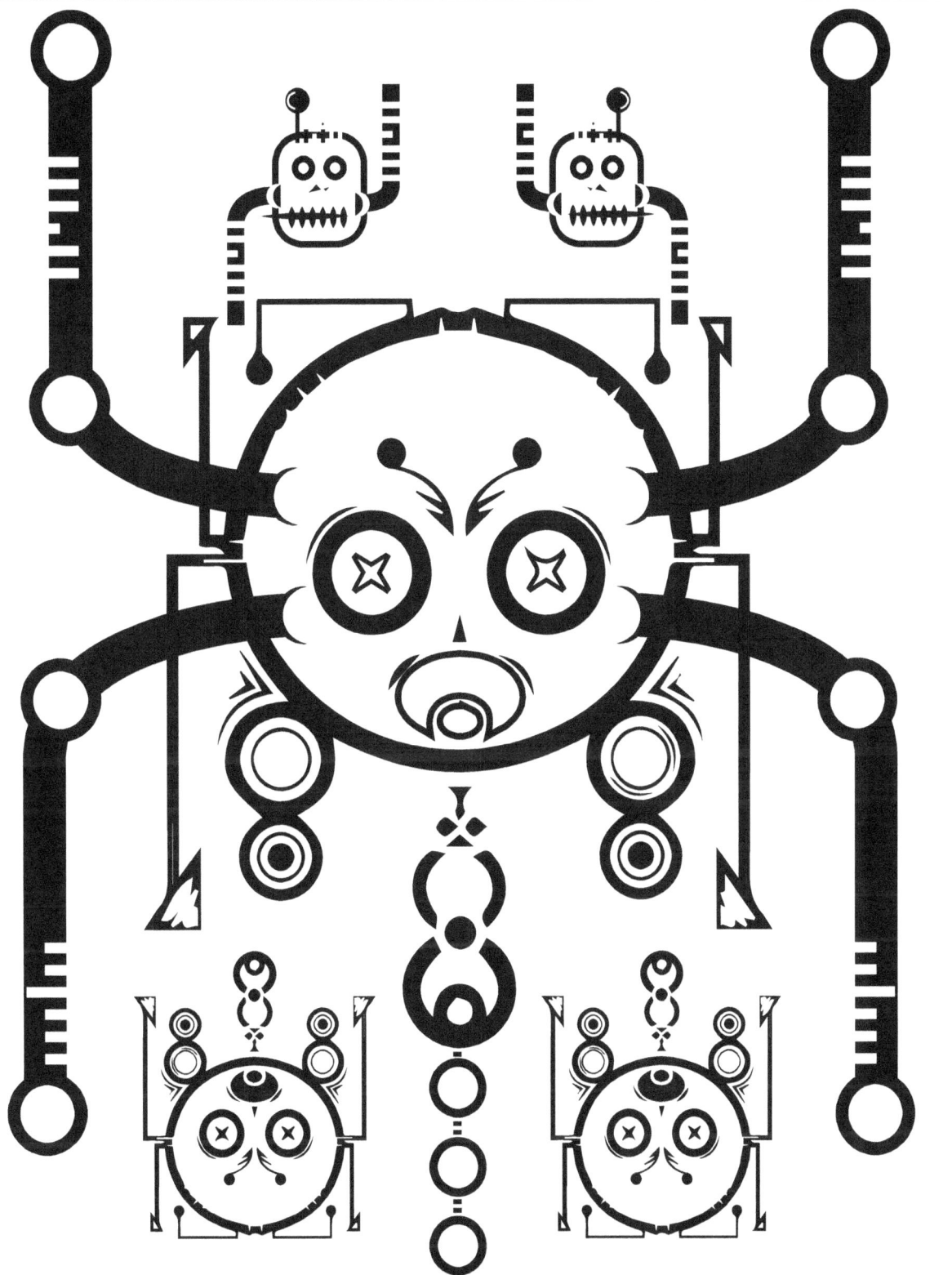

METRIC TON 19.

doodles by drewski

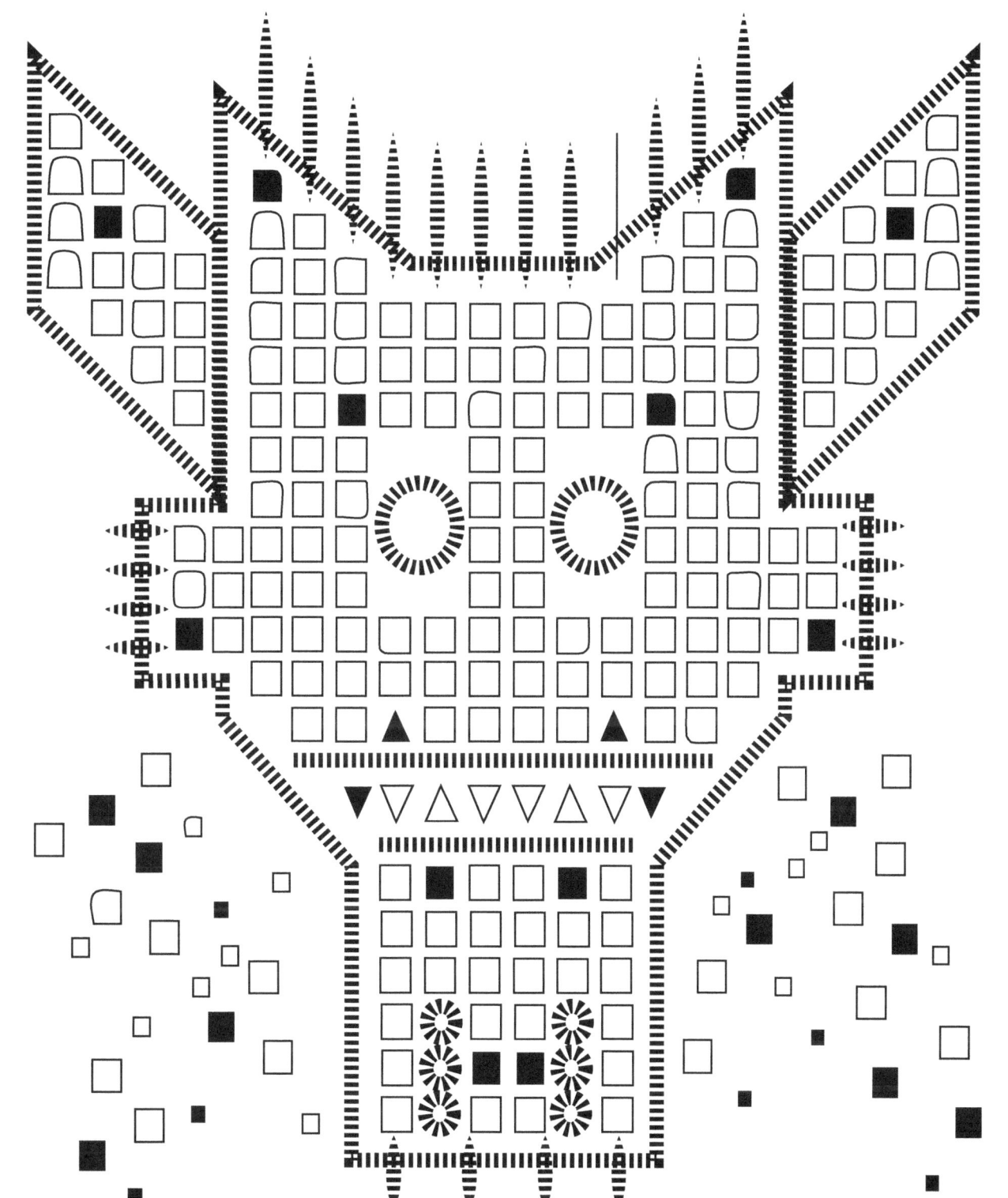

PIXEL HATE

doodles by drewski

doodles by drewski